Carlos E. Baldelomar
Karina M. Vilchez
Yaoska S. Méndez

Neural Networks

Carlos E. Baldelomar
Karina M. Vilchez
Yaoska S. Méndez

Neural Networks

In Credit Risk Management

ScienciaScripts

Imprint

Cover image: www.ingimage.com

This book is a translation from the original published under ISBN 978-620-0-01193-0.

Publisher:
Sciencia Scripts
is a trademark of
Dodo Books Indian Ocean Ltd. and OmniScriptum S.R.L publishing group

120 High Road, East Finchley, London, N2 9ED, United Kingdom
Str. Armeneasca 28/1, office 1, Chisinau MD-2012, Republic of Moldova, Europe
Managing Directors: Ieva Konstantinova, Victoria Ursu
info@omniscriptum.com

Printed at: see last page
ISBN: 978-620-8-62596-2

Table of Contents

Foreword

In the age of information and digital transformation, financial risk management faces unprecedented challenges and opportunities. Neural networks, as a fundamental part of the development of artificial intelligence, have emerged as one of the most powerful tools to address the inherent complexity of credit risk assessment. This book, "Neural Networks in Credit Risk Management," aims to provide an overview of how these technologies are transforming the financial industry, improving the ability of institutions to assess risk more accurately and effectively.

The book begins with an accessible and well-grounded introduction to the basic concepts of artificial neural networks, including the structure of input, hidden, and output layers, as well as how learning algorithms work. This conceptual foundation is essential for readers to understand how neural networks learn to identify complex patterns in large volumes of data, a key process in credit risk assessment.

The history and evolution of neural networks is also presented in detail, highlighting the milestones that have allowed these technologies to evolve from simple mathematical models to robust prediction and analysis tools. From the pioneering ideas of Warren McCulloch and Walter Pitts to the advances in backpropagation led by David Rumelhart and Geoffrey Hinton, the authors take us through a historical journey that contextualizes the current impact of neural networks in the financial sector.

One of the main contributions of this book is its focus on the practical applications of neural networks within the financial industry, particularly in credit management. It explores various network architectures, such as feedforward networks and recurrent networks, and discusses how each can be useful in solving specific problems in the field credit. The ability of these networks to learn from historical data and adapt to new circumstances is essential to improve the accuracy in the evaluation of credit applicants, allowing greater efficiency in decision making and minimizing the risk of default.

The authors also address the challenges and limitations in implementing these technologies, recognizing the complexity of the models and the importance of high-quality, representative data. The "black box" nature of neural networks and the risks of bias in the data are discussed in depth, offering recommendations on explainability and transparency techniques that can help overcome these obstacles and ensure that decisions made by the models are ethical and understandable.

This book not only provides a technical overview of how neural networks work, but also illustrates how their application can transform critical processes in the financial industry, such as credit risk assessment, fraud detection and investment portfolio optimization. The cases and examples presented clearly show how these tools enable financial institutions to adapt to a competitive and constantly changing environment, improving both operational efficiency and the quality of service offered to customers.

In short, "Neural Networks in Credit Risk Management" presents itself as essential reading for academics, financial professionals, and anyone interested in understanding how artificial intelligence technologies are redefining risk analysis and decision making in finance. This book offers a unique combination of theory and practice, inviting readers to explore the potential of neural networks to improve financial management more efficiently, accurately, and equitably.

Chapter 1 : Introduction to Neural Networks and their Application to Credit Management

What are neural networks: basic concepts?

Artificial neural networks (ANNs) are computational models inspired by the functioning of the human brain. They consist of layers of interconnected nodes, also called "neurons", which work collaboratively to learn complex patterns from data. These networks are organized in three main types of layers: the input layer, which receives the initial data; the hidden layers, where the complex processing is performed; and the output layer, which provides the final results (Del Carpio Gallegos, 2005), in **Figure 1**.

Figure 1: Layers of Neural Networks

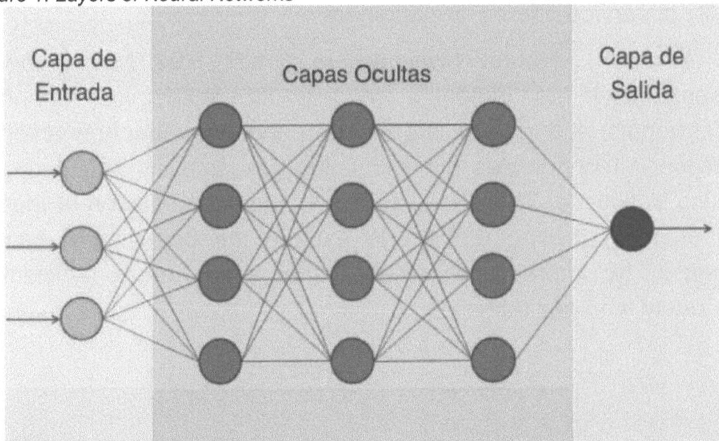

Source: https://aprendeia.com/que-son-las-redes-neuronales-artificiales/

Input layer

The input layer is the first layer of the neural network and its function is to receive the raw data to be processed. Each node in this layer represents a specific characteristic of the input data set. The input layer does not perform any complex processing; it simply distributes the received information to the following layers of the network for further analysis.

Hidden layers

Hidden layers are those located between the input layer and the output layer. These layers are responsible for information processing. Each node in a hidden layer receives signals from the nodes in the previous layer, applies a weight and an activation function as shown in **Figure 2** (such as the sigmoid, ReLU or tanh function) and transmits the result to the next layer. The hidden layers allow the network to capture complex patterns and nonlinear relationships present in the data, which is critical for classification and prediction tasks.

Figure 2: Sigmoid function, ReLU or tanh

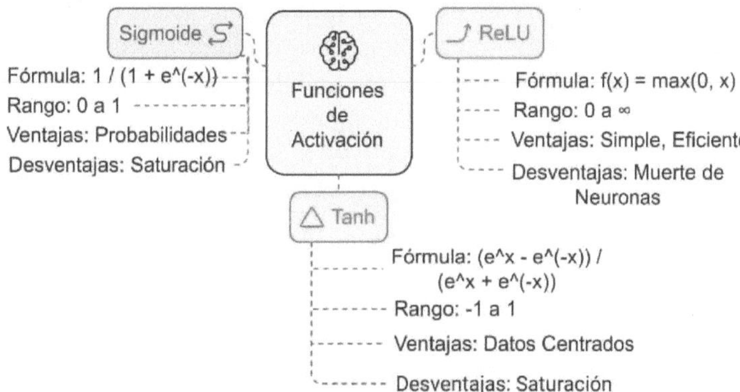

Source: Own elaboration

Output layer

The output layer is the final layer of the neural network. Its function is to produce the output or prediction based on the processing performed by the hidden layers. Each node in the output layer represents a possible output class or value. Depending on the nature of the problem, the output layer may have a single neuron (in regression problems) or multiple neurons (in classification problems). This process of forward propagation and iterative adjustment of weights allows neural networks to "learn" and generalize from examples. This mechanism emulates the synaptic transmission between biological neurons, allowing neural networks to perform classification, regression and prediction tasks with great effectiveness (Aldabas-Rubira, 2002).

The learning of neural networks is based on the iterative adjustment of the weights of the connections between nodes through a process called training. During training, a data set called training set is used, which allows the network to adjust its internal parameters in order to minimize the error in its predictions. This process involves multiple iterations in which the network compares its output with the actual values, calculates the error and adjusts the connection weights to reduce the discrepancy.

Error backpropagation is one of the most important algorithms in this learning process as seen in **Figure 3**. This algorithm calculates how the errors are distributed backward through the network, and then uses these calculations to update the weights of each connection. The backpropagation algorithm uses the gradient of the error with respect to each weight to determine the direction and magnitude of the necessary adjustment. Optimization techniques, such as stochastic gradient descent, are employed to make these adjustments, allowing the network to find a configuration of weights that minimizes the total error. This technique is especially efficient in multilayer networks, where interactions between multiple hidden layers allow complex patterns to be modeled.

Figure 3: Backpropagation Process in Neural Networks

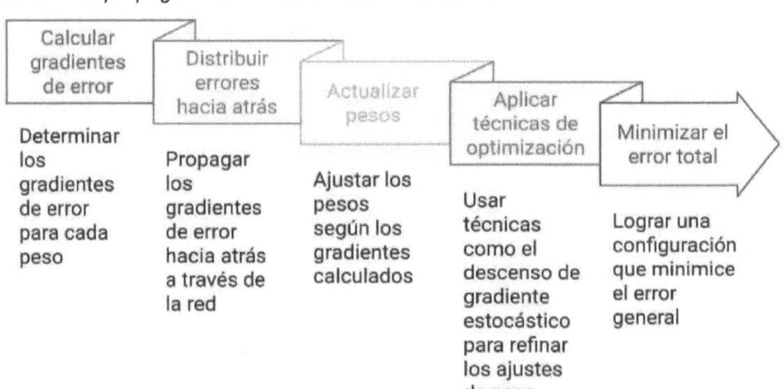

Source: Own elaboration

Thanks to this iterative, weight-based fitting process, neural networks can learn and discover complex, nonlinear relationships between input variables. This makes them particularly useful for a wide variety of tasks that require the detection of subtle patterns in large amounts of data, such as credit risk prediction, where the probability of default of a credit applicant is assessed, image recognition, where specific features in photographs or videos need to be identified, and natural language processing, which involves understanding and generating human text effectively.

Neural network models are capable of nonlinear pattern recognition, which makes them ideal for tackling complex problems in finance, including credit risk management. These capabilities not only enable the development of robust models, but also provide the flexibility to generalize from historical data and adapt to changing scenarios, which is crucial for coping with the dynamic nature of financial markets. Neural networks, unlike traditional statistical models, can identify non-obvious relationships and adapt to unexpected changes in data patterns, resulting in greater accuracy and adaptability.

Neural networks are especially useful for predicting financial behaviors, such as the probability of default, and classifying credit applicants into risk categories, which facilitates decision making based on real data (Toro Ocampo, Mejía Giraldo & Salazar Isaza, 2004). For example, by analyzing the profiles of credit applicants, these networks can combine complex variables such as credit history, income and demographic characteristics to generate a more accurate risk score. This allows financial institutions to optimize credit granting and reduce the probability of delinquency.

According to Méndez Araya (2009), neural networks represent a set of interconnected processing elements that mimic the knowledge acquisition process of the human brain. This parallelism with the functioning of the human brain makes them highly effective for prediction in complex financial environments, since they are capable of learning and adapting their parameters based on the experience acquired during training. In this way, neural networks improve the accuracy and reliability of their estimates over time, which is essential for credit risk management, where the ability to respond to new market conditions is fundamental (Méndez Araya, 2009). In addition, continuous learning allows them to update their parameters when new data is presented, which provides a significant advantage in a changing and competitive financial environment.

History and evolution of neural networks

The concept of artificial neural networks was first proposed in the 1940s by Warren McCulloch and Walter Pitts, who developed a mathematical model describing the functioning of biological neurons as logical devices capable of processing input signals (McCulloch & Pitts, 1943). This mathematical approach made it possible to conceptualize neurons as binary units that are activated or inhibited depending on certain thresholds, laying the foundation for the development of the artificial neural networks we know today. McCulloch and Pitts' idea was revolutionary, since it proposed that complex cognitive processes could be represented by combinations of simple logical operations, which was fundamental for the advancement of artificial intelligence.

Later, in 1949, Donald Hebb introduced a theory of learning known as Hebb's rule, which described how neurons could modify their connections as a function of synaptic activity (Hebb, 1949). Hebb proposed that the connection between two neurons is strengthened if both are activated simultaneously, an idea synthesized in the phrase "neurons that fire together, fire together." This principle forms the basis of adaptive learning in modern neural networks, as it allows connections between neurons to adjust according to experience, facilitating the formation of complex patterns. Hebb's theory had a profound impact on neuroscience and artificial intelligence, as it introduced the concept of synaptic plasticity, which is essential for learning and memory (Ruiz & Basualdo, 2001).

In the 1980s, a significant breakthrough came with the introduction of the error backpropagation algorithm, initially developed by Paul Werbos and later popularized by David Rumelhart, Geoffrey Hinton and Ronald Williams, which enabled the practical application of multilayer networks to solve complex supervised learning problems (Werbos, 1974; Rumelhart, Hinton & Williams, 1986). Error backpropagation is a key algorithm that allows to adjust connection weights efficiently, minimizing the error of network predictions through the use of gradient descent. This advance was instrumental in overcoming the limitations of early perceptron models, which could not solve nonlinearly separable problems.

Thanks to backpropagation, multilayer neural networks, also known as multilayer perceptron (MLP) networks, became much more powerful and capable of modeling complex relationships between inputs and outputs. This advance not only enabled the resolution of more sophisticated classification problems, but also facilitated data prediction in areas such as speech recognition, computer vision, and financial risk management. The availability of powerful computing systems, such as graphics processing units (GPUs), and access to large volumes of data over the last few decades have led to the resurgence and success of neural networks in numerous fields, including finance (Ramírez & Chacón, 2011).

The use of neural networks in finance has been particularly beneficial for risk assessment and market behavior prediction, thanks to their ability to learn from large volumes of historical data and detect complex patterns that would be difficult to identify using traditional methods. This resurgence has also been driven by increasing data storage capacity and the development of more efficient optimization algorithms, which have made the training of deep neural networks faster and more effective.

General applications of neural networks in industry

Neural networks have been successfully employed in various

industries because of their ability to model and solve complex problems by learning hidden patterns in data. Thanks to their structure and ability to learn from examples, neural networks can identify non-obvious features and capture non-linear relationships that would be difficult to detect using traditional methods. This enables them to offer innovative solutions in a variety of industries and contexts.

In the health sector, for example, they are used in the diagnosis of diseases, taking advantage of their ability to analyze medical images and other clinical data with great precision. Neural networks have proven to be especially useful in detecting abnormalities in X-rays, MRI and CT scans, enabling healthcare professionals to identify early signs of pathologies such as cancer or cardiovascular disease. In addition, they are used in the analysis of electronic health records, helping physicians to make better informed decisions based on historical patterns and patient data (Ramírez & Chacón, 2011).

In the retail industry, these networks make it possible to predict consumer behavior and personalize offers for each customer. Through the analysis of purchase data, browsing history and individual preferences, neural networks can segment customers and anticipate their needs, improving the efficiency of marketing strategies and increasing loyalty. For example, recommendation systems used by large e-commerce platforms rely on neural networks to suggest products that match the particular interests and needs of each user, which increases conversion rates and improves customer experience (Aldabas-Rubira, 2002).

In addition, in the industrial sector, neural networks have been applied to predictive maintenance of machinery. By analyzing historical data from sensors and operational logs, these networks can predict failures or breakdowns before they occur, allowing companies to reduce downtime, optimize the performance of their operations, and reduce costs associated with reactive maintenance. This predictive approach is critical for industries that rely on operational continuity, as it enables efficient maintenance planning and minimizes unexpected outages.

In the financial field, neural networks have proven to be particularly useful for fraud detection, asset price prediction, credit management and portfolio optimization. In fraud detection, neural networks are able to analyze patterns of financial behavior and transactions in real time, identifying suspicious activities that could be indicative of fraud. Thanks to their ability to learn and adapt to new fraudulent tactics, these networks are a powerful tool for ensuring the financial security of institutions and customers. In the field of asset price prediction, neural networks are used to analyze time series and detect historical patterns in prices, facilitating the prediction of future market movements, which helps investors make more informed decisions and minimize risks.

In credit management, neural networks are used to assess the risk profile of applicants, considering multiple socioeconomic and financial variables to determine the probability of default. These models allow for a comprehensive analysis of each applicant, incorporating information such as credit history, income, demographic characteristics, and past payment behavior. This comprehensive assessment generates a more accurate risk score than traditional methods, which helps to optimize credit granting, minimize delinquency losses and offer better credit conditions to each client, adjusted to their specific risk profile, as shown in **Figure 4.**

Figure 4: Factors Contributing to Credit Risk Assessment

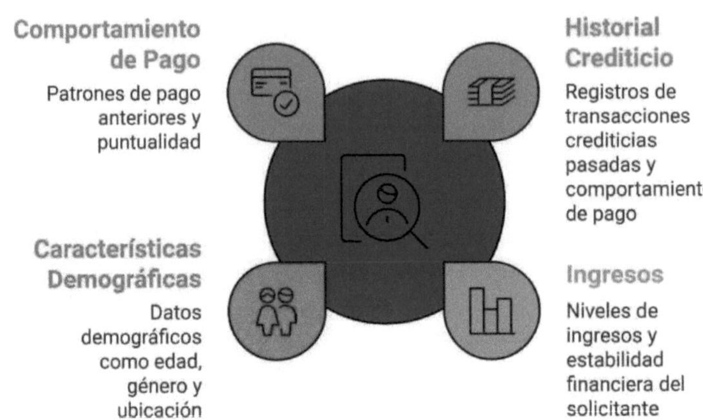

Source: Own elaboration

Likewise, neural networks have begun to be used in the optimization of investment portfolios. These models allow the simultaneous analysis of a large number of assets and their correlations to build a portfolio that maximizes return and minimizes risk. By analyzing historical data and identifying patterns, neural networks are able to generate more adaptive investment strategies, adjusting to market changes and improving portfolio performance.

According to Guerrero et al. (2024), the incorporation of artificial intelligence, including neural networks, has made it possible to automate many financial tasks, improving both the accuracy and speed of decision making (Guerrero et al., 2024). In the field of credit risk assessment, the use of neural networks has facilitated the creation of more accurate and adaptive models that allow a large number of variables to be analyzed simultaneously. These variables can include both traditional financial aspects, such as payment history and debt-to-income ratio, as well as additional, less conventional factors, such as social networking behavior or consumption patterns.

This comprehensive approach allows financial institutions to generate much more detailed and specific risk profiles for each applicant, which in turn leads to more informed and personalized decisions. For example, neural networks can identify complex patterns that indicate a high probability of default, even when certain traditional financial indicators do not appear to show risk. This provides a significant competitive advantage by anticipating potential default problems more effectively.

In addition, the continuous learning capability of neural networks allows models to automatically update themselves when new data is incorporated, thus ensuring that risk assessments remain accurate over time. This adaptability is crucial in a dynamic financial environment, where market conditions and consumer behaviors can change rapidly. As a result, financial institutions can reduce delinquency rates and optimize their loan portfolio, offering suitable conditions for each risk profile in an efficient manner.

This has not only improved the operational efficiency of financial institutions, but has also increased the quality of the services offered, allowing clients to obtain faster and more reliable responses in critical processes, such as credit risk assessment and investment management.

Credit management: challenges and opportunities

The main objective of credit management is to assess an applicant's ability to meet the payment obligations arising from a loan. This process is essential to mitigate financial risk and ensure the sustainability of financial institutions. A proper credit risk assessment not only allows institutions to protect themselves against potential losses, but also to optimize their loan portfolio, thus contributing to the stability and profitability of the financial system as a whole.

Traditionally, credit evaluation has been based on classical statistical models that apply rigid rules to rank applicants, such as financial ratio analysis or the calculation of credit scores using logistic regression algorithms. These approaches are useful when structured data and linear relationships between variables are available. However, they have significant limitations, especially when dealing with large volumes of complex and unstructured data, such as transactional data, historical records or qualitative characteristics that are difficult to quantify (Toro Ocampo et al., 2004).

Neural networks offer a powerful alternative to overcome these challenges. Thanks to their ability to learn hidden patterns and nonlinear relationships in data, neural networks can make more accurate predictions even in complex and changing environments. One of the main advantages of these models is their ability to handle and process large volumes of data, making it possible to extract valuable information from a variety of heterogeneous data sources, integrating both traditional financial variables and unstructured data. This data can include consumer behavior, social media interactions, credit history, recent transactions and even demographic characteristics that traditional statistical models do not usually incorporate.

This comprehensive approach significantly improves the predictive ability of models, as it takes into account a wide variety of factors that contribute to an applicant's likelihood of default. For example, while traditional models focus primarily on financial history and credit score, neural networks can consider behavioral patterns in credit card usage, spending on non-essential goods, and even frequency of payments on other types of services. This provides a more complete and contextual view of each applicant's risk profile , helping financial institutions make more informed and targeted decisions.

In addition, neural networks can dynamically update their models as new data is received, which is crucial to reflect the evolution of customers' financial behavior and adjust quickly to changing market conditions. This ensures that credit risk assessments are always current and relevant, reducing the likelihood of errors in applicant classification. In a financial environment where changes can occur suddenly, the ability of neural networks to continuously adapt provides a considerable competitive advantage. Furthermore, the automation of these processes helps to reduce the turnaround time in credit evaluation, offering customers a faster and more efficient experience.

These capabilities allow not only to improve credit rating, but also to detect possible fraud and anomalous behavior, which is key in risk management (Van Greuning, 2009). Neural networks can analyze large volumes of transactions in real time, detecting suspicious patterns that could indicate fraudulent activities, such as unusual transactions or inconsistencies in user behavior. This early detection capability is essential to minimize financial losses and protect both institutions and their customers.

In addition, neural networks are well suited to the current demands of the financial industry, which requires systems capable of making fast and accurate decisions to provide a competitive advantage. The dynamic nature of the financial industry requires models to be able to update and adjust quickly to market changes and customer behavior. The continuous learning capability of neural networks allows these models to improve their accuracy as new data is collected, keeping assessments always relevant and adapted to current circumstances.

For example, when new types of financial products are introduced or macroeconomic conditions change, neural networks can recalibrate their models to reflect these changes, thus ensuring that risk assessments are accurate and up-to-date. This results in reduced default rates, as credit granting decisions are better informed and aligned with each applicant's current risk profile. At the same time, the optimization of financial resources is benefited, as they are allocated more efficiently to those clients with an appropriate profile, which improves the overall profitability of financial institutions (Ramírez & Chacón, 2011).

Neural networks offer a powerful alternative to overcome these challenges, as they have the ability to learn hidden patterns in data and make accurate predictions even in complex and changing environments.

In addition, neural networks enable a substantial improvement in credit risk assessment by incorporating data from multiple sources, including traditional financial data, credit history, and other more complex indicators, such as social networking behavior and consumption patterns. By combining and analyzing these factors, neural networks can generate more accurate risk scores than traditional methods, resulting in a fairer and more informed assessment of credit applicants. This ability to integrate heterogeneous information is particularly valuable today, where customers have increasingly diverse and complex financial profiles.

The opportunities offered by neural networks in the financial industry are not limited to risk assessment and fraud detection, but also include the optimization of operational processes. Neural networks are well suited to the current demands of the industry , which requires systems capable of making fast and accurate decisions, offering a significant competitive advantage. The continuous learning capability of neural networks allows models to be dynamically updated with new data, ensuring that risk assessments remain relevant and adjusted to changing market conditions (Ramírez & Chacón, 2011).

However, implementing neural networks also presents significant challenges. These include the inherent complexity of the models, which can be difficult to interpret, affecting transparency in decision making. This has raised regulatory concerns, as financial institutions must be able to explain how and why a particular decision was made. The "black box" nature of neural networks poses a challenge in terms of trust and regulatory compliance, particularly in situations where a credit rejection needs to be justified. In addition, there is a risk of introducing bias into the models if the training data is unrepresentative or biased, which can lead to discriminatory decisions.

Despite these challenges, the opportunities offered by neural networks in credit risk management are considerable. Their ability to process large volumes of data, identify complex patterns, and adapt models in real time provides financial institutions with a powerful tool to improve the quality of their decisions, reduce risk, and optimize their operations. The key to seizing these opportunities lies in balancing technological innovation with measures to ensure transparency, fairness and regulatory compliance, thus ensuring that neural networks are used responsibly and ethically.

The impact of artificial intelligence on financial decision making

The incorporation of artificial intelligence (AI) and, in particular, neural networks, has significantly transformed the financial sector. AI facilitates the automation of complex processes, such as risk assessment, credit application approval, and investment portfolio management, improving the efficiency, accuracy and speed of decisions (Ruiz & Basualdo, 2001). Neural networks, by learning complex patterns from large volumes of historical data, provide a solid foundation for informed decision making. These capabilities make it possible to anticipate potential problems, such as the probability of loan default or the detection of fraudulent behavior, which would be difficult to identify using traditional methods (Sosa Sierra, 2007).

The impact of AI extends to multiple aspects of the financial industry, most notably in the personalization of services and improved risk management. In credit risk assessment, neural networks make it possible to simultaneously analyze a wide variety of data, from traditional financial information to alternative data such as social network behavior and consumption patterns. This makes it possible to generate more robust and accurate models that are better adapted to the changing profiles of consumers, and that offer a more detailed and contextual view of each credit applicant. In addition, the ability of neural networks to continuously learn and adjust their models with new data makes them extremely useful in a dynamic financial environment where macroeconomic conditions and customer behaviors can change rapidly.

AI has also boosted the operational efficiency of financial

institutions by automating previously manual and time-consuming tasks, such as sorting credit applications, analyzing financial documents, and managing customer service . Thanks to advances in natural language processing (NLP), chatbots and virtual assistants are now able to provide immediate and personalized customer support, answering frequent queries and streamlining procedures. This not only improves the customer experience, but also frees up human resources that can be redirected to more complex and value-added tasks.

In portfolio optimization, neural networks help identify complex correlations between different assets and generate investment strategies that maximize return and minimize risk. This adaptive approach makes models more resilient to unexpected market fluctuations, providing investors with advanced tools for managing their assets. By analyzing large volumes of historical and macroeconomic data, neural networks can detect emerging patterns that indicate trend changes, allowing fund managers to proactively adjust their strategies.

However, the implementation of neural networks in the financial sector is not without its challenges. One of the main challenges is the lack of transparency, as these models operate as a "black box", making it difficult to interpret the results and explain the decisions made. In many cases, even developers and subject matter experts have difficulty understanding how a neural network has reached a specific conclusion due to the complexity of the interactions between its multiple layers and weights. This is particularly problematic in the financial sector, where clarity and accountability are essential, especially when it comes to decisions that directly affect consumers, such as the approval or rejection of a credit application. Lack of explainability can generate distrust among both clients and regulators, as those affected need to understand why they are being denied or approved for credit.

Another major challenge is the risk of algorithmic bias. If the data used to train the network contains historical biases, these can be amplified by the network, leading to unfair or discriminatory decisions. For example, if certain demographic groups have had less access to credit in the past, the model could learn this pattern and perpetuate it, automatically and unintentionally discriminating against those groups. This problem is particularly relevant in the financial context, where equitable access to credit is crucial for social and economic inclusion.

In addition, the technical infrastructure required to train and operate large-scale neural networks represents another challenge. Deep neural networks require enormous data processing and storage capacity, which implies a significant investment in technological infrastructure. This need for computational resources may be an obstacle for some financial institutions, especially smaller ones that do not have access to advanced technological resources.

On the other hand, the implementation of neural networks also faces challenges related to regulatory compliance. Regulatory environments in the financial sector are strict and tend to evolve slowly compared to the rapid development of technology. This creates a disconnect between the capacity for innovation offered by neural networks and the constraints imposed by existing regulations. Financial institutions must ensure that the use of these technologies complies with all current regulations, which may require considerable effort in terms of documentation, testing and justification of the models used.

To overcome these challenges, it is crucial to adopt approaches that ensure transparency and fairness in the use of neural networks. This includes the development of explainability techniques that enable neural network models to provide more understandable interpretations of how they arrive at their decisions. Tools such as LIME (Local Interpretable Model-agnostic Explanations) and SHAP (SHapley Additive exPlanations) are increasingly being used to provide local explanations of predictions, helping users and regulators understand the factors that influenced a specific decision. In addition, special attention should be paid to the quality and representativeness of training data to mitigate the risk of bias, as well as periodic audits to assess the impact of models on different demographic groups.

Despite these challenges, the benefits that neural networks bring to the financial sector are considerable, provided they are implemented responsibly and ethically. The key to seizing these opportunities lies in balancing technological innovation with measures that ensure transparency, fairness and regulatory compliance, thus ensuring that neural networks are used fairly and effectively, to the benefit of both financial institutions and consumers.

Chapter 2 : Technical Fundamentals of Neural Networks for Credit Management

Structure and operation of a neural network

Artificial neural networks (ANNs) have transformed credit risk assessment through their ability to learn and model complex patterns in large volumes of data. In the context of credit management, the operation of a neural network is centered on the process of machine learning, which allows models to continuously adjust and improve as they are exposed to more historical and new data. This section describes the functioning of the layers of a neural network, highlighting its impact on credit risk management and its ability to optimize efficiency and accuracy in financial decision making (Del Carpio Gallegos, 2005).

Input layer

In a model for credit risk management, the neurons in the input layer could represent variables such as the applicant's credit history, monthly income, age, education level, occupation, and other relevant demographic characteristics (Perales Paz, 2024).

The correct configuration of the input layer is crucial for the performance of the neural network, since an inadequate selection of variables can negatively affect the network's ability to learn useful patterns. For this reason, data preprocessing techniques, such as normalization and standardization, are often applied before input to the input layer in order to improve the quality of learning and ensure that all variables contribute adequately to the model. The quality of the data supplied to the input layer will largely determine the effectiveness of the model in predicting credit risk.

Hidden layers

In the context of credit risk management, hidden layers allow for in-depth analysis of how different financial and personal variables of credit applicants interact. For example, these layers can detect subtle patterns in customers' payment behaviors that could indicate a higher probability of default. Furthermore, thanks to the use of algorithms such as error backpropagation, the network adjusts its weights to minimize the difference between the predicted output and the actual value, which improves the accuracy of the model over time (Pérez Ramírez & Fernández Castaño, 2007).

Output layer

In a financial context, the output layer may have a single neuron if a continuous prediction is being made, such as estimating a credit score. Alternatively, it may have multiple neurons if the goal is to classify the applicant into different risk categories, such as "low risk," "medium risk," or "high risk" (Melchor Perez et al., 2024). The output of the network is used for decision making, such as approving or rejecting a credit application, based on the evaluation of the applicant's profile.

The joint operation of these three layers allows neural networks to be a powerful tool for credit risk management. The input layer facilitates the collection of relevant data, while the hidden layers enable deep analysis of the nonlinear interactions between these variables, capturing complex patterns and hidden features that would not be evident with traditional methods. Finally, the output layer synthesizes all the processed information to produce a prediction or classification that is easily interpretable by financial decision systems.

This integrated process allows neural networks to analyze large volumes of data and detect complex relationships between multiple variables, providing financial institutions with the ability to make more accurate and informed predictions. For example, the networks can identify profiles of applicants who have a high probability of meeting their financial obligations or those who present a higher risk of default. This predictive capability is essential not only to minimize financial risk, but also to optimize credit allocation, allocating resources efficiently and improving the profitability of the loan portfolio. Moreover, by continuously learning from historical and new data, neural networks adapt to changing market conditions, allowing financial institutions to maintain up-to-date and accurate assessment models.

Types of neural networks used in credit management

There are several types of neural networks used in credit management, each with specific capabilities for solving complex problems, as shown in **Figure 5.**

Figure 5: Types of Neural Networks

Redes Neuronales en la Gestión de Crédito

Redes Neuronales Feedforward (FNN)	Redes Neuronales Recurrentes (RNN)	Redes Neuronales Convolucionales (CNN)	Redes Neuronales de Retroalimentación	Redes Generativas Antagónicas (GAN)
Clasificación y Regresión	Análisis de Datos Secuenciales	Detección de Patrones Complejos	Evaluación de Secuencias Financieras	Generación de Datos Sintéticos

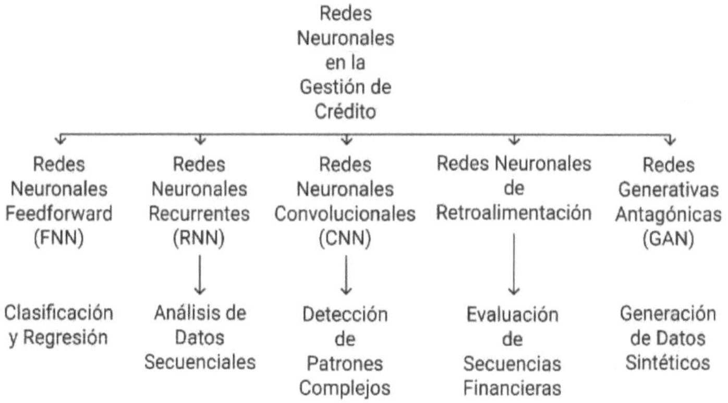

Source: Own elaboration

The main types of neural networks applied in the credit context:

Feedforward Neural Networks (FNN)

Feedforward neural networks are the most basic and are used for classification and regression tasks in the field of credit risk. These networks process information in one direction only, from the input layer to the output layer, without cycles or feedback. They are useful for predicting the probability of default of a credit applicant, classifying it as low, medium or high risk. Due to their simple structure, feedforward networks are easy to train and offer good interpretability, which is valuable for justifying decisions to financial regulators.

Recurrent Neural Networks (RNN)

Recurrent neural networks are suitable for the analysis of sequential and time series data. In credit management, RNNs are useful for analyzing a customer's financial behavior over time, considering historical patterns of payments, transactions and other sequential events. Their architecture allows information to persist in the network, making predictions dependent on past history. For example, an RNN can help identify whether a customer with a history of on-time payments is showing signs of possible default due to recent changes in financial behavior. However, a limitation of traditional RNNs is the gradient fading problem, which makes it difficult to learn long-term dependencies. To overcome this, variants such as LSTM (Long Short-Term Memory) and GRU (Gated Recurrent Unit) are used, which improve the ability of RNNs to capture long-term patterns in the data.

Convolutional Neural Networks (CNN)

Although convolutional neural networks are predominantly used in image recognition and spatial data processing, they can also be adapted for credit management. In the financial domain, CNNs can be useful for detecting complex patterns in large volumes of high-dimensional data. For example, a CNN can be applied to identify complex relationships between various characteristics of credit applicants that are not evident through traditional analysis. CNNs are able to capture correlations between different variables, providing unique insights that can improve the accuracy of predictive models.

Feedback Neural Networks (Feedback Neural Networks)

Feedback networks, also known as recurrent neural networks

in some contexts, have connections that allow the output of certain neurons to be reintroduced as input to the same previous layer or layers. This feedback causes the network to have a temporal "memory," which is useful for assessing the sequence of financial events in a customer's history. In credit risk management, this ability to remember previous patterns is particularly valuable for predicting future behavior based on past actions.

Generative Adversarial Neural Networks (GAN)

GANs are a type of neural network consisting of two competing networks: a generator network and a discriminator network. Although GANs are not used directly to classify credit risks, they can be used to generate synthetic data to complement training data sets. In the context of credit management, this can be useful when little labeled data is available, allowing the training quality of other neural network models to be improved.

Each of these types of networks has specific applications in credit management, and the selection of the appropriate type depends on the nature of the data and the specific problem to be solved. Combining different types of networks is also a common practice to take advantage of the strengths of each to improve the accuracy and robustness of predictions.

Supervised and unsupervised learning algorithms

In the context of credit risk management, supervised and unsupervised learning algorithms play a crucial role in improving the accuracy of predictive models and providing a more detailed assessment of applicants' risk profiles. The rationale and applications of both types of learning in the field of credit management are discussed below.

Supervised Learning

Supervised learning is a type of algorithm in which the model learns from a labeled data set, i.e., each training example is accompanied by a desired outcome or label. In the credit domain, this could mean that the dataset contains information from past applicants along with a label indicating whether the credit was successfully repaid or if there was a default. Through this labeled data, the model learns to establish relationships between applicant characteristics (e.g., income, age, credit history) and the desired outcome (repayment or default).

Some of the most common supervised learning algorithms used in credit management include the following:

- **Logistic Regression**: It is one of the most widely used methods in credit risk assessment due to its simplicity and ability to interpret linear relationships between variables. Although it is simpler than other methods, it is still an effective tool for classifying applicants as high or low risk. In addition, logistic regression allows for the calculation of probabilities, which is very useful in determining the degree of risk associated with each applicant, thus providing a quantitative basis for decision making. This interpretability is key for financial institutions to justify their decisions to regulators and ensure transparency in credit evaluation processes.
. **Decision Trees and Random Forest**: These algorithms create a model based on a series of decision rules, which are useful for identifying non-linear relationships in the data. Decision Trees allow the visualization of decisions and their possible consequences, which facilitates the interpretation and justification of the decisions made. Random Forest, in particular, is a set of decision trees that allow to improve accuracy by using the average of multiple trees, thus reducing the risk of overfitting. In addition, Random Forest is less sensitive to outliers and offers greater robustness to noisy data, which is essential in credit risk assessment where data often

have variability.

- **Neural Networks**: Using a feedforward architecture, neural networks are also applied in supervised learning to predict the probability of non-compliance. These networks allow the integration of a large number of variables and capture complex patterns that cannot be detected by simpler methods (Montalván, 2019). In addition, neural networks have the ability to learn nonlinear representations of data, which is especially useful when there are complex interactions between applicant characteristics. This allows them to provide more accurate predictions compared to traditional models, thus providing a significant advantage in credit risk assessment.

The use of supervised learning facilitates the construction of accurate models that can predict the probability of default based on historical patterns. However, one of the challenges with this approach is the need for a large amount of labeled data, which can be costly and difficult to obtain, especially when recent data representative of current market behavior is required (Perales Paz, 2024).

Unsupervised Learning

Unsupervised learning, unlike supervised learning, does not require labeled data. Instead of learning a relationship between input variables and a known label, unsupervised learning seeks to identify hidden patterns and structures in the data without prior information about the desired outcomes. In the field of credit management, this type of learning can be used to segment credit applicants into different groups based on behavioral similarities and demographic characteristics.

Some of the most widely used unsupervised learning algorithms in credit management include:

- **Clustering**: Algorithms such as K-means or DBSCAN allow grouping credit applicants into segments according to common characteristics. For example, groups of applicants with similar profiles can be identified, which is useful for offering customized financial products or identifying potential risks. K-means is effective for grouping large volumes of data into homogeneous clusters, while DBSCAN is particularly useful for detecting patterns in the presence of noise or outliers, which is crucial in credit management where data quality can be variable. In addition, the use of clustering allows the identification of subgroups that present specific risks, facilitating proactive and personalized risk management.

- **Principal Component Analysis (PCA)**: This dimensionality reduction technique allows simplifying large data sets by eliminating redundancy, which facilitates the identification of important patterns without losing too much information. In credit management, PCA is used to reduce the number of variables and make the model more efficient. In addition, PCA helps to improve the speed of model training and reduce the risk of overfitting, since working with a smaller number of relevant variables minimizes model complexity and avoids overfitting the model to the training data. This is particularly valuable in credit management, where large volumes of data are handled and it is essential to identify the most influential characteristics for decision making.

Unsupervised learning is particularly valuable when working with unstructured data or when it is not known a priori which characteristics are the most relevant for analysis. In credit risk management, it allows the discovery of emerging groups or patterns that might not have been considered previously, which can significantly improve customer segmentation and offer personalization (Hernández and Vásquez, 2023). In addition, by identifying hidden patterns in data, unsupervised learning facilitates the identification of new risk factors that may not be evident to the naked eye, allowing for more proactive and accurate risk assessment. This helps financial institutions develop adaptive strategies for different customer profiles and improve the effectiveness of risk mitigation policies.

Comparison and Combined Application

Although supervised and unsupervised learning are used for different purposes, both can be complementary in the context of credit management. For example, unsupervised learning can first be used to segment customers into groups based on common characteristics, and then supervised learning can be used to build specific predictive models for each group. This combination improves the accuracy of the models and makes it possible to offer financial products that are better suited to the needs and profiles of each customer segment.

The use of these algorithms in credit management has allowed financial institutions to improve both the accuracy and efficiency of their decision making. However, the correct implementation of these methods requires an adequate technological infrastructure and access to quality data, as well as a thorough understanding of the limitations and potential biases that the models may introduce. The combination of both types of learning, together with advanced data preprocessing techniques, constitutes one of the most effective strategies to optimize credit risk assessment and ensure responsible management of financial resources (Fernandez and Torres, 2024).

Data preprocessing and its importance in risk prediction.

Data preprocessing is a critical step in building machine learning models, particularly in credit risk management, where data quality has a direct impact on the accuracy of predictions. Before training a model, the data must be cleaned, transformed and normalized to ensure that the model can learn from it as effectively as possible. Below are some of the key processes involved in data preprocessing and their importance in credit risk prediction.

Data Cleansing

Data cleaning is one of the first and most important steps in preprocessing. Credit application data may contain missing values, duplicates, typographical errors or inconsistencies in formats, which can negatively affect model performance. To address this problem, techniques such as missing value imputation (which can be performed using statistical methods or predictive models), duplicate removal and typo correction are used. In addition, integrity checks can be applied to ensure consistency between different data sources. These steps are essential to ensure that the model is trained with information that is accurate and representative of reality, thus reducing the possibility of bias and improving the overall quality of the model (Garcia Lopez, 2024).

Normalization and Standardization

The data used in credit risk models often have different scales. For example, an applicant's annual income may range from thousands to millions, while his or her age is measured in tens. To prevent characteristics with larger values from dominating the training process, normalization and standardization techniques are applied to bring all values to a comparable scale. This improves the efficiency of the model and allows relevant features to have a fair weight in predictive decisions (Rodriguez and Martinez, 2023).

Transformation of Characteristics

In some cases, the original variables must be transformed so that they can be used effectively by machine learning algorithms. For example, categorical features such as marital status or occupation of the applicant must be converted to a numerical format using techniques such as one-hot coding. In addition, new features can be created from existing ones (feature engineering) to provide the model with more relevant information. This process helps to highlight patterns that might not be evident in the original features and, therefore, improves the model's ability to predict risks (Hernández and Vásquez, 2023).

Data Imbalance Management

In credit risk management, the data set may be unbalanced, i.e., the number of applicants who have defaulted in the past is usually much smaller than the number of those who have complied with their payments. This imbalance can bias the model towards the majority class, making it less effective in identifying potential defaulters. To address this problem, techniques such as oversampling the minority class, undersampling the majority class, or using specific algorithms that account for the imbalance are used, which helps improve the model's ability to correctly identify high-risk customers (Fernandez and Torres, 2024).

Importance of Preprocessing in Risk Prediction

Data preprocessing not only improves the quality of data used by models, but also contributes significantly to the accuracy and robustness of predictions. In the context of credit risk, good preprocessing can make the difference between a model that properly identifies applicants at high risk of default and one that produces biased or incorrect results. In addition, by standardizing

data and manage imbalances, preprocessing ensures that the model is fairer and more equitable, minimizing the risk of discrimination or erroneous decisions that could have significant financial consequences for both the institution and the applicants (Perales Paz, 2024).

Preprocessing also facilitates the reduction of data dimensionality, which is crucial for optimizing model performance and avoiding overfitting. The removal of redundant or irrelevant features allows the model to focus on the variables that actually provide predictive value, which improves both the accuracy and speed of the training process. In addition, by ensuring data quality and consistency, preprocessing contributes to greater model interpretability, allowing financial analysts and decision makers to better understand the reasons behind the predictions and act accordingly. This not only improves confidence in the model, but also enables better communication of the results to stakeholders, including regulators and clients.

Evaluation of neural network models in the context of credit.

The evaluation of neural network models in the context of credit risk is a crucial step to ensure that the predictions made by the models are accurate, reliable and applicable in real financial environments. Some of the most relevant approaches and metrics for evaluating the effectiveness of neural network models in credit risk management are described below.

Evaluation Metrics

Evaluation metrics allow measuring the performance of neural network models in ranking credit applicants. Common metrics include:

. **Accuracy**: Accuracy measures the ratio of correct predictions to the total predictions made. While accuracy is a useful metric, in the context of credit risk it can be misleading if the data set is unbalanced, as a model that always predicts the majority class could have high accuracy without being useful. In addition, accuracy does not provide information on the types of errors made, such as false positives or false negatives, which are especially relevant in credit risk analysis, as each type of error has different financial and operational implications for the institution.

. **Confusion Matrix**: The confusion matrix is a tool to evaluate model performance in terms of true positives, true negatives, false positives and false negatives. This helps to better understand the errors made by the model and to identify possible areas for improvement. In addition, the confusion matrix provides a detailed view of how the model classifies different cases, allowing specific analysis of the type of errors that occur most frequently. This is critical in the lending environment, where minimizing false positives and false negatives has a direct impact on the efficiency and profitability of the financial institution.

. **Precision and Recall**: Precision indicates how many of the cases predicted as positive actually are, while recall measures how many of the actual positive cases were correctly identified by the model. In the context of credit risk, these metrics help to assess how many high-risk clients were correctly identified, as well as the potential cost of not identifying some of them. A high accuracy value reduces the number of false positives, which is crucial to avoid granting loans to clients who will not be able to meet their obligations. On the other hand, a high recall value ensures that most high-risk customers are identified, minimizing the risk of undetected defaults. These metrics, when used together, allow achieving the right balance between minimizing credit risk and maximizing the chances of granting credit to good applicants.

. **ROC curve and AUC (Area Under the Curve)**: The ROC curve shows the relationship between the true positive rate and the false positive rate at different decision thresholds. The AUC (Area Under the Curve) measures the ability of the model to distinguish between positive and negative classes. A high AUC value indicates that the model has a good ability to discriminate between applicants who meet their payments and those who do not.

- **Fl-Score**: The F1-score is the harmonic mean between precision and recall. This metric is especially useful when there is a significant imbalance between classes, as it provides a more balanced assessment of model performance.

Figure 6: Credit Risk Model Evaluation

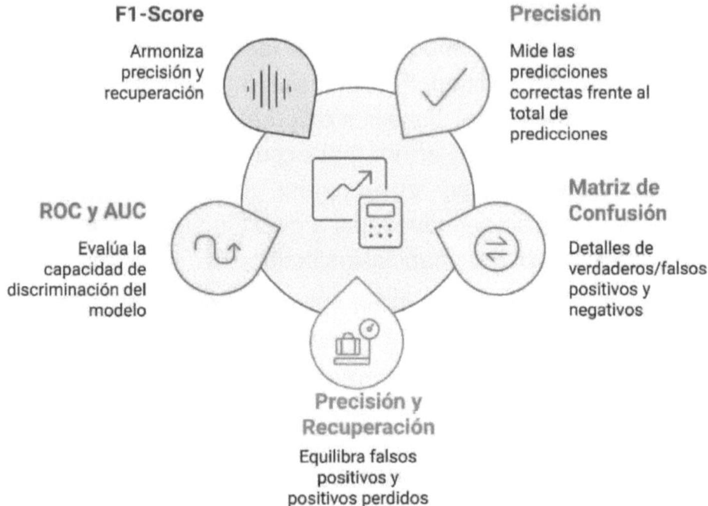

Source: Own elaboration

Cross Validation

Cross-validation is a technique used to evaluate the robustness of the model and its ability to generalize to new data. In credit risk management, cross-validation is used to divide the data set into multiple partitions, training the model on some of them and evaluating it on others. This ensures that the performance of the model does not depend solely on a specific partition of the data, thus improving confidence in its generalizability.

Importance of Interpretability

In the financial context, model interpretability is a key aspect in the evaluation of neural networks. Financial institutions must be able to explain why a model has made a certain decision, especially when a credit is denied. To improve interpretability, techniques such as LIME (Local Interpretable Model-agnostic Explanations) and SHAP (SHapley Additive exPlanations) can be used to analyze the impact of each feature on model predictions. This not only facilitates regulatory compliance, but also improves the confidence of clients and other stakeholders in the transparency of the credit evaluation process.

Stress and Robustness Tests

Neural network models used in credit risk management must be evaluated not only in terms of accuracy and performance, but also in terms of their robustness to adverse situations. Stress testing consists of subjecting the model to extreme scenarios, such as economic crises or abrupt changes in applicant profiles, to verify how it behaves under unfavorable conditions. These tests are essential to ensure that the model is able to maintain acceptable performance even under unforeseen circumstances, thus guaranteeing the stability of the financial institution.

Cost of Classification Errors

In the evaluation of neural network models for credit risk, it is important to consider the cost associated with classification errors. Not all errors have the same impact: a false positive (approving a loan to a client who cannot pay) has different financial consequences than a false negative (rejecting a client who could have met his obligations). Therefore, the evaluation should include a cost analysis to minimize the financial impact of errors made by the model and ensure optimal decision making for the institution.

The comprehensive evaluation of neural network models in the credit context ensures that they are accurate, interpretable, robust and efficient in risk prediction. This not only improves the quality of credit decisions, but also contributes to the financial stability of the institution and a better customer experience.

Chapter 3: Implementation and Challenges in the Use of Neural Networks for Credit Management

How to implement a neural network in credit systems

The implementation of neural networks in credit systems requires a combination of technical and practical expertise, as well as an adequate infrastructure. This includes both the hardware needed to handle large volumes of data and the software platforms to efficiently implement and optimize the models. In addition, it is critical to have a multidisciplinary team that includes data science experts, software engineers, and credit specialists to ensure that all stages of the process are adequately addressed as shown in the figure below.

Figure 7: Implementation of Neural Networks in Credit Systems

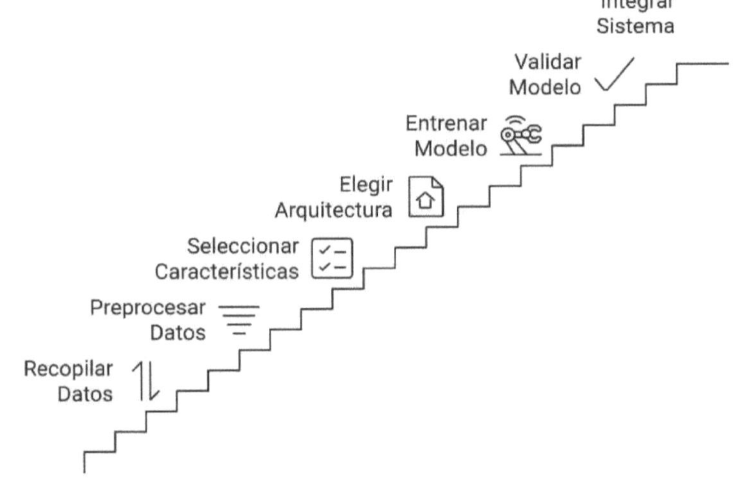

Source: Own elaboration

To begin with, the problem to be solved must be clearly defined, whether it is the prediction of defaults, customer classification, risk estimation or even the identification of anomalous behavior patterns. This definition must involve both technical and financial domain experts to ensure that the objectives of the model are consistent with the needs of the business.

Next, it is essential to collect historical data on credit applicants, which will serve as the basis for training the model. These data should be comprehensive and include financial, demographic, and behavioral characteristics, such as income, credit history, payment patterns, level of indebtedness, and previous relationships with the financial institution (Villamil Bahamón, 2013). It is also important to consider the inclusion of alternative data, such as social networking behavior or utility usage patterns, that can complement the traditional credit risk assessment.

The data collection process must follow strict quality protocols to ensure that the information is reliable and representative. This involves the elimination of outliers or inconsistent values, imputation of missing values, and standardization of variables to ensure consistency between the different data sets used.

The first step is to preprocess the data, ensuring that it is clean, normalized and free of inconsistent values. This process involves several key steps, such as imputation of missing values, removal of duplicate data, and transformation of categorical variables into formats suitable for processing by the neural network, such as one-hot coding. It is also important to normalize the data to ensure that all variables are on the same scale, which facilitates model training and prevents some features from dominating over others due to their numerical rank.

Subsequently, a feature selection must be performed to reduce the dimensionality of the data set and keep only the most relevant variables. This may involve the use of techniques such as Principal Component Analysis (PCA) or selection based on feature importance. The quality of the preprocessing has a significant impact on the accuracy and generalizability of the model.

Once the preprocessing is completed, the most appropriate neural network architecture is chosen according to the problem: a feedforward network for basic classification problems, or a recurrent network to capture temporal patterns in payment behavior data. In some cases, convolutional neural networks (CNN) can be used to extract complex features, especially when the input data has some spatial structure, as can be the analysis of scanned financial documents (Basogain Olabe, 2014). Architecture selection is a critical step, as it must be aligned with the type of data and the objective of the analysis to maximize the efficiency and accuracy of the model.

Once the architecture is designed, the model is trained using optimization algorithms, such as gradient descent, and cost functions, such as cross-entropy, to adjust the network weights. Training involves feeding data to the model and adjusting the weights at each iteration to minimize the difference between predictions and actual values. This process is done through multiple epochs, which allow the model to learn complex patterns over time.

The implementation also requires the selection of optimal hyperparameters, such as the number of hidden layers, the number of neurons per layer, the learning rate, the batch size, and the number of epochs. The choice of these hyperparameters is crucial to achieve good model performance. For example, too high a learning rate may lead to the model not converging correctly, while too low a rate may result in excessively slow training.

Another important aspect is the use of regularization techniques, such as L2 regularization or dropout, to prevent overfitting, which occurs when the model learns the details and noise of the training data too well and thus loses generalization ability. L2 regularization adds a penalty term to the total cost of the model that depends on the magnitude of the weights, which helps to keep the weights under control and avoids excessive model complexity. On the other hand, dropout involves the random deactivation of neurons during training, which forces the model to be more robust and less dependent on specific neurons.

The use of cross-validation is also critical to evaluate model performance and iteratively adjust the hyperparameters to obtain the best possible performance without incurring overfitting. In particular, k-fold cross-validation is a widely used technique, where the data set is divided into k subsets, and the model is trained k times, using a different subset for validation each time. This allows a more robust evaluation to be obtained and ensures that the model performs well on different splits of the data set.

In addition to these techniques, methods such as early stopping, which stops training when the performance on the validation set stops improving, can be implemented, thus preventing the model from continuing to adjust to noise in the data. All these techniques combined allow building a model that not only performs well on training data, but also generalizes well to new data, which is crucial in the context of credit risk management.

The training process can also benefit from the use of advanced techniques such as dynamic learning rate adjustment, using algorithms such as Adam (Adaptive Moment Estimation) or RMSprop (Root Mean Square Propagation). These algorithms adjust the learning rate during the training process, which allows the model to converge faster and avoid being trapped in local minima. Adam combines the advantages of gradient descent with momentum and learning rate adaptivity, while RMSprop adapts the step size based on the mean of recent gradients, which is particularly useful in problems with high variability.

In addition, proper initialization of weights is essential to accelerate convergence and avoid problems such as fading or gradient explosion, especially in deep networks. Techniques such as Xavier initialization or He initialization allow to distribute the weights in an optimal way, favoring a constant flow of gradients along the network.

Constant monitoring of performance metrics, such as loss and accuracy, during training allows problems to be identified early and necessary adjustments to the architecture or hyperparameters to be made. This also includes monitoring validation metrics to detect if the model is starting to overfit the training data. By proactively identifying these problems, strategies such as modifying the learning rate, adjusting the number of hidden layers or applying additional regularization techniques can be implemented to improve overall model performance (Pérez Ramírez & Fernández Castaño, 2007).

Finally, the trained model must be integrated into the credit management system, ensuring that it can be accessed in real time to make decisions on credit applications. This integration must be robust and ensure data security, as financial information is highly sensitive (Ladino Becerra, 2014). It is essential that the integration be done with operational efficiency and system scalability in mind, allowing the model to handle an increasing volume of applications without loss of performance.

To ensure successful integration, several best practices should be taken into account, as shown in the figure:

Figure 8: Successful Model Integration

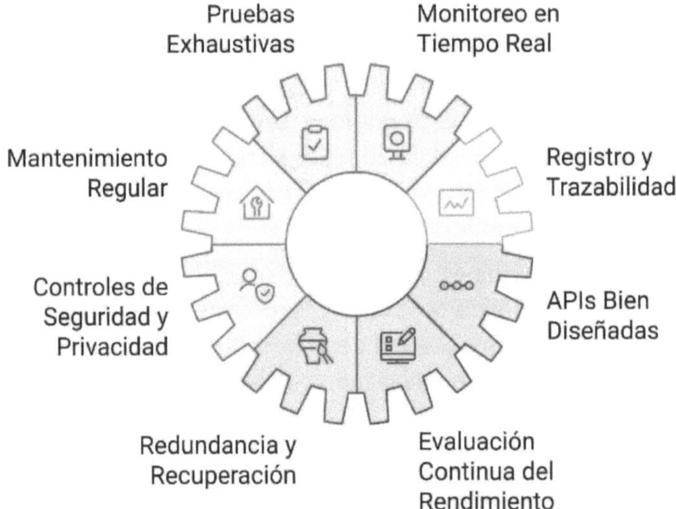

Source: Own elaboration

- **Extensive integration testing**: Perform extensive testing to ensure that the model will function correctly when integrated with existing systems. This includes stress testing to ensure that the system can handle high volumes of requests.
- **Real-time monitoring**: Implement monitoring tools to verify the performance of the model in production. This will help to quickly identify any problems or anomalies that may arise during the use of the model in real situations.

- **Regular maintenance and updating**: Plan for regular updating of the model to ensure that it continues to be relevant in the face of changes in applicant behavior or market conditions.
- **Log management and traceability**: Implement a detailed logging system for all predictions made by the model. This not only helps in monitoring performance, but also provides essential traceability in case of audits or further analysis.
- **Security and privacy controls**: Ensure that the system complies with all security and data privacy requirements, including encryption of sensitive information and strong authentication to prevent unauthorized access.
- **Well-designed application programming interfaces (APIs)**: Use well-designed and documented APIs to integrate the model efficiently with other systems in the credit workflow, ensuring smooth communication and reducing the possibility of errors.
- **Redundancy and failover**: Implement redundancy and failover mechanisms to ensure that the system is resilient and can continue to function even in the event of technical problems.
- **Continuous performance evaluation**: Maintain a constant cycle of model performance evaluation using accuracy, recall, and F1-score metrics. This will allow timely adjustments to the model to maintain its effectiveness.

Challenges and limitations in the implementation of neural networks

The implementation of neural networks in credit management is not without its challenges and limitations, many of which affect both the feasibility and effectiveness of these models. One of the main challenges is data quality and quantity. Neural networks require large volumes of data to be properly trained, and this data must be of high quality to ensure model accuracy. For neural network models to be effective, it is critical to have data that accurately represents the characteristics and behaviors of credit applicants. Data quality implies not only the absence of errors, but also the relevance and timeliness of the information used. In the credit context, collecting reliable and complete data can be a challenge, as it often relies on multiple sources that may have inconsistencies, missing data, or even biases inherent to the collection systems (Villamil Bahamón, 2013).

In addition, the data must be sufficiently diverse to capture a wide variety of loan applicant profiles. Lack of diversity in the training data can lead to models that do not generalize well, especially when faced with loan applications from populations that were not well represented in the initial data. For example, if the training data is biased toward a specific group of applicants, such as those with high incomes or residents of urban areas, the resulting model may not be able to correctly evaluate lower-income or rural applicants. This results in potentially unfair and discriminatory financial decisions.

This challenge is compounded when there are restrictions on the availability of historical data, either due to lack of digitization, the fragmented nature of financial data, or lack of access to certain data segments due to privacy policies. In addition, data for certain minority or marginalized groups may be unavailable or insufficient, which increases the risk of the model developing systematic bias. Data diversity should also consider factors such as gender, ethnicity, geographic location, and socioeconomic status to ensure that the model has an inclusive and equitable approach to credit risk assessment.

To address this problem, it is crucial to implement more inclusive data collection strategies that consider the diversity of applicant profiles. This may include collaboration with community-based organizations and the use of alternative data that can better represent underrepresented populations. In addition, data augmentation techniques, which involve generating synthetic data to balance differences in training samples, can be used to reduce bias and improve the model's ability to generalize to different applicant profiles.

The integration of alternative data sources, such as information from social networks, consumption patterns, or non-traditional financial behaviors, could improve data quality and diversity. Social networks, for example, can provide real-time information on applicants' behavior and socioeconomic status, such as changes in employment, recent activities, or consumption patterns. However, the use of these data presents considerable privacy-related challenges, as the collection and use of data from social networks can violate users' confidentiality if not properly managed. In addition, there are concerns about the validity of this data, as information on social networks may be manipulated or may not accurately represent an individual's actual financial behavior.

Another major challenge is the consistency and standardization of data from multiple sources. Alternative sources, such as mobile transaction data, IoT (Internet of Things) sensor data, or even data on online browsing behavior, can provide valuable insights into the behavioral patterns of credit applicants. However, this data is often in widely varying formats and must be unified in order to be used effectively in predictive modeling. This requires an intensive process of data transformation and cleansing to ensure that all information is compatible and aligned with the objectives of the model.

In addition, alternative data must also be validated to ensure reliability and to avoid introducing noise into the model, which could decrease its performance. These challenges must be addressed by rigorous preprocessing and validation of the data, ensuring that each source is reliable and contributes positively to model performance. The use of cross-validation techniques and the involvement of human experts to verify the quality of the collected data are essential strategies to mitigate the risks associated with the integration of these new sources.

A significant challenge related to the implementation of neural networks is the computational complexity, especially in the case of deep networks with multiple hidden layers. Training these networks can be extremely costly in terms of time and computational resources, as they require a large amount of processing to tune the millions of parameters involved. This represents a considerable barrier for small or medium-sized financial institutions that do not have the necessary technological infrastructure, such as servers with GPUs or access to cloud computing platforms, to train and maintain these models efficiently (Basogain Olabe, 2014).

The need for specialized hardware, such as graphics processing units (GPUs) or even tensor processing units (TPUs), is an additional challenge that increases implementation costs. In addition, the lack of access to cloud computing platforms that can provide these resources on demand limits the ability of many organizations to experiment with and scale the use of neural networks in credit risk assessment. This technological barrier is also reflected in the lack of trained personnel to manage these infrastructures and optimize the models, which adds an additional difficulty.

To overcome these challenges, it is essential to consider strategies such as the use of lighter models or learning transfer techniques to reduce computational requirements. In addition, collaboration with cloud service providers can help institutions access the necessary resources in a more cost-effective and flexible manner. However, these solutions require careful planning to ensure the efficiency and long-term sustainability of the infrastructure used.

A critical challenge in using neural networks is to ensure that the model does not become overly dependent on the data it was trained on, a phenomenon known as overfitting. Because of the ability of neural networks to learn complex patterns, there is a significant risk that the model will learn specific details of the training data too well, including noise and peculiarities that are not generalizable. As a result, when the model is used with new data, it may fail to recognize different patterns, leading to inaccurate predictions.

For example, if a model is trained only on loan applicant data from a particular economic context, such as a period of stability, it may have difficulty making accurate predictions in different situations, such as a recession. This overfitting phenomenon occurs because the model becomes extremely good at predicting the data set it was trained on, but loses flexibility to adapt to new contexts.

To mitigate this problem, techniques such as regularization, which penalizes excessive model complexity by adding penalty terms in the cost function, and cross-validation, which allows the performance of the model to be evaluated on different subsets of the data to ensure that it generalizes well, are used. In addition, techniques such as dropout, which randomly disables some neurons during training, can help improve the model's ability to generalize by reducing the risk that it learns specific patterns from the training data. However, these approaches are not always sufficient, especially if the training data are limited or lack the necessary diversity, which can compromise the robustness of the model (Pérez Ramírez & Fernández Castaño, 2007).

Another related problem is the opacity of neural network models. These models are considered "black boxes", as the complexity of their internal layers makes it difficult to interpret how a particular decision is arrived at. In the field of credit management, this represents a major problem, since financial decisions must be explainable and transparent, both for applicants and regulators (Ladino Becerra, 2014). This lack of interpretability can generate distrust both at the institutional level and for clients, which can hinder the adoption of these technologies.

Finally, there are regulatory and ethical barriers. The adoption of neural networks in credit management involves dealing with sensitive applicant data, which is subject to privacy and data protection regulations. In addition, the possibility of bias in training data can lead to discriminatory results, which has legal and ethical implications. Ensuring compliance with these regulations and ensuring fair and equitable treatment of all applicants are challenges that must be carefully addressed. For example, mechanisms should be implemented to audit the model for potential biases and establish mitigation protocols to correct any identified problems (Herrera Ortega, 2023).

Ethical and regulatory considerations in the use of artificial intelligence for lending.

The application of neural networks in credit risk assessment raises a number of ethical and regulatory considerations that must be addressed to ensure that these technologies are used in a fair and responsible manner. These considerations span different key areas that need to be carefully analyzed and managed.

Data privacy

Privacy of applicant data is a key concern. Neural network models require large volumes of personal data to train, which poses risks related to information protection. It is necessary to ensure that the data collected is protected against unauthorized access and used only for the stipulated purposes. This involves complying with regulations such as the General Data Protection Regulation (GDPR) in the European Union, which establishes strict rules on the collection, storage and use of personal data (Herrera Ortega, 2023).

In addition, it is important to ensure that data collection is done under the explicit consent of applicants, who must be fully informed about how their data will be used and the rights they have over it, such as the right to request deletion of their information. This approach is not only necessary for regulatory compliance, but also strengthens trust between clients and financial institutions.

The use of alternative data, such as information from social networks or consumer transactions, presents specific privacy challenges. While this data can provide valuable information to assess credit risk in a more comprehensive manner, its collection must be done with special care to protect the privacy of individuals. Financial institutions should implement strict policies on the use of this data, ensuring that any use is aligned with applicants' privacy expectations and complies with applicable regulations. In addition, it is important to keep in mind that this data may be particularly sensitive and may not always reflect an accurate financial context, requiring responsible and transparent treatment.

Implementing advanced anonymization and encryption techniques is another crucial practice for protecting data privacy. Anonymization ensures that personal data cannot be traced back to specific individuals, while encryption protects information during transmission and storage. These practices are critical to minimize the risks of data breaches and ensure responsible handling of personal information.

Fairness in credit decisions

Fairness in credit decisions is a key consideration, especially given the potential for neural networks to perpetuate or amplify biases present in training data. These models, being complex, can learn biased patterns if the data reflect existing inequalities in society, such as differences in treatment of certain groups based on gender, ethnicity, or socioeconomic status. This could lead to discriminatory decisions, where certain groups of applicants are systematically disadvantaged in credit evaluation processes, affecting both their access to financial services and their economic well-being.

To mitigate this risk, it is essential to implement model auditing and control mechanisms. These mechanisms should allow for the detection and correction of possible biases during the training and production phase of the model. For example, a recommended practice is to periodically evaluate model performance in different segments of the population to identify if there are disparities in pass rates among specific groups. This constant evaluation allows for parameter adjustment and improved model fairness.

In addition, it is essential to ensure that models are auditable so that both applicants and regulators can have clarity on how decisions are made. Transparency in the development and evaluation of models helps to identify potential biases, which facilitates the building of trust between financial institutions and their clients. Implementing techniques to evaluate the importance of the characteristics used by the model can help ensure that there are no variables that implicitly introduce discrimination (Herrera Ortega, 2023).

Another relevant strategy is the use of bias mitigation approaches during data preprocessing, such as collecting balanced data or adjusting weights to make underrepresented groups more visible. Techniques can also be employed at the training stage, such as penalizing biased results, to improve fairness. Implementing these strategies contributes to the creation of a fairer credit evaluation system, where all applicants have an equal opportunity to be evaluated on their merits and not because of biases in the data.

Model explainability

Model explainability is a key component to ensure transparency and confidence in the decisions made by neural networks. In the financial field, the ability to explain how and why a certain decision was made is crucial, not only to comply with regulations, but also to maintain the confidence of credit applicants and financial institutions. Unlike traditional models, such as those based on linear regression, neural networks are often considered "black boxes", as the complexity of their internal structure makes it difficult to interpret the decision process.

To address this lack of interpretability, it is important to implement methods to unravel the logic behind the predictions. Tools such as LIME (Local Interpretable Model-agnostic Explanations) and SHAP (SHapley Additive exPlanations) are useful in this context, as they allow identifying which specific features most influenced a particular decision in the model. These tools offer a way to provide understandable explanations that can be presented to both regulators and applicants, thus increasing the level of transparency of the process (Ribeiro, Singh, & Guestrin, 2016; Lundberg & Lee, 2017).

The use of these tools not only facilitates the understanding of individual decisions, but also helps to identify general patterns in model behavior that might indicate the presence of biases. For example, if certain attributes, such as place of residence or gender, are found to have a disproportionate influence on model decisions, it is possible to intervene to adjust the parameters and ensure a fairer assessment.

In addition, the development of explainable models also involves working on simplifying network architectures where possible, without compromising the accuracy of predictions. This may involve reducing the number of hidden layers or using hybrid architectures that combine elements of more transparent models with neural networks to improve the explanatory power of the decisions made.

It is essential to conclude that improving the explainability of neural networks is essential for building trust and ensuring regulatory compliance. This is especially important in the financial arena, where institutions must justify their decisions to regulators and credit applicants. The use of tools such as LIME (Local Interpretable Model-agnostic Explanations) and SHAP (SHapley Additive exPlanations) can help provide understandable explanations of the decisions made by the model, thus facilitating a higher degree of transparency (Ribeiro, Singh, & Guestrin, 2016; Lundberg & Lee, 2017).

Human supervision and accountability

Human supervision plays a crucial role in the implementation of neural networks for credit risk assessment. While artificial intelligence can automate many processes, human intervention remains indispensable to ensure fair and ethical decisions. Human operators must verify that the model results are aligned with the principles of fairness and transparency, acting as a control mechanism to detect possible errors (Herrera Ortega, 2023).

One of the key roles of human supervision is to review cases where the model generates questionable or atypical results. These cases can have a significant impact on the financial well-being of applicants, and human intervention allows for an assessment of whether there are additional factors that the model has not adequately considered. This ensures that all decisions are balanced and fair, avoiding negative consequences for certain groups (Ribeiro, Singh, & Guestrin, 2016).

In addition, human oversight is key to ensuring regulatory compliance. Decisions made by automated systems must comply with laws and regulations, such as data protection regulations and guidelines on fair credit assessment practices. Human operators must be responsible for ensuring that decisions are aligned with legal standards and for intervening when the model deviates from these (Lundberg & Lee, 2017).

Operator training is equally important. In order for operators to intervene effectively, they must understand both the capabilities and limitations of neural network models. This includes understanding the potential biases that may be present in the data and the performance of the algorithms, which will enable them to intervene in an informed and proactive manner when necessary.

Finally, accountability and human oversight remain essential. Although artificial intelligence models can automate much of the credit evaluation process, final decisions must be overseen by humans to ensure that all relevant factors are being considered and acted upon fairly. Operators must be trained to understand the limitations of the models and act accordingly, intervening when necessary to correct errors or unfair decisions.

Future of neural networks in the financial industry

The future of neural networks in the financial industry is promising, with numerous opportunities and challenges that will define their impact on credit risk management and other financial services. As technology advances and larger volumes of data accumulate, neural networks are expected to play an increasingly important role in financial analysis and decision making.

Expansion of applications and improvement of model accuracy

One of the main trends for the future is the expansion of neural network applications in the financial sector. Today, these networks are already used for credit risk assessment, fraud detection, and customer segmentation, but in the coming years they could also be applied in areas such as market prediction, asset management, and the automation of complex financial processes. The ability of neural networks to analyze unstructured data, such as comments on social networks and other types of alternative data, will enable a more accurate and contextual assessment of the financial profile of credit applicants.

In addition, steady improvement in model accuracy is expected due to advances in the development of more sophisticated architectures and access to more diverse and extensive data sets. Hybrid architectures, which combine neural networks with other algorithmic approaches, are also gaining ground. These models have the potential to improve predictive capability by taking advantage of different techniques, such as supervised learning and unsupervised deep learning.

Reducing bias and improving equity

The future will also need to address the ethical challenges currently facing neural networks in the financial context. Reducing bias and improving model fairness will continue to be critical issues. As more companies adopt this technology, developers are expected to focus on implementing best practices to mitigate the biases inherent in training data and ensure that the decisions made by the models are fair to all applicants. The development of stricter regulatory frameworks could be an essential tool to ensure fairness and transparency in the use of these systems (Herrera Ortega, 2023).

To address the problem of bias, a key strategy will be the integration of explainability techniques, such as LIME and SHAP, that allow developers and stakeholders to better understand how predictions are generated. These tools will not only facilitate the auditability of models, but will also help identify areas where decisions may be being unfair, allowing for adjustments before models are implemented in the operational environment (Ribeiro, Singh, & Guestrin, 2016; Lundberg & Lee, 2017).

Advances in Interpretability and Widespread Adoption

Another relevant aspect for the future is the development of more interpretable models. As regulatory pressure and customer demand for transparency increases, it will be essential for neural networks to be able to explain their decisions in a way that is understandable to everyone involved. It is expected that more advanced methods will be developed to interpret how neural networks process data and make decisions, making it possible for even non-technical users to understand and trust the model results.

Interpretability will be a key factor in the widespread adoption of neural networks by financial institutions. Those organizations that are able to implement interpretable models and meet transparency expectations will have a significant competitive advantage, as they will be able to establish greater trust with their customers and comply with regulations more effectively. In addition, collaboration between regulators, financial institutions and technology experts will be critical to fostering the safe and ethical adoption of neural networks in the financial sector.

Technical challenges and future opportunities

Despite the significant opportunities, the use of neural networks in the financial industry brings with it several technical challenges that must be addressed. One of the biggest challenges is the computational complexity involved in training deep models, which demands large processing and storage capacity. This implies the need for specialized infrastructure, such as graphics processing units (GPUs) and tensor processing units (TPUs), which allow model training to be performed more efficiently and quickly. These infrastructures can be expensive and not always accessible to all institutions, especially smaller ones, which creates a significant barrier to the adoption of these technologies. However, advances in cloud computing platforms are helping to democratize access to these resources, offering more affordable and scalable solutions (Lundberg & Lee, 2017).

Another relevant technical challenge is the need to improve the efficiency of model training to reduce the associated time and costs. Currently, the training process can be extremely intensive, both in terms of time and resources. In this context, the development of techniques such as federated learning could revolutionize the industry. Federated learning allows models to be trained using distributed data without the need to centralize the information, which not only improves efficiency, but also offers significant benefits in terms of data privacy and security, which is critical for the financial industry.

In addition, managing unstructured data presents a significant challenge. Neural network models have the potential to leverage valuable information from non-traditional sources, such as social networks or browsing histories. However, processing and effectively utilizing this data is complex, as it requires considerable preprocessing to be properly integrated into the model. Advances in natural language processing (NLP) and sentiment analysis techniques will be critical for this data to be effectively integrated, providing a more complete and accurate view of applicants' credit profiles.

Finally, the integration of new technologies, such as quantum computing, could offer innovative solutions to scalability and efficiency problems in the future. Quantum computing has the potential to significantly reduce training time and improve the ability of models to analyze large volumes of data. Although this technology is still under development, its advancement could radically change the landscape of neural networks in the financial industry, offering opportunities to solve some of the most complex technical challenges facing the sector.

Collaboration between regulators and developers

Finally, the future of neural networks in the financial industry will depend to a large extent on close collaboration between technology developers, financial institutions and regulators. The deep integration of these technologies into financial decision making will require clear standards to govern their use, thereby promoting ethics and safety in their implementation. This collaboration will allow developers to understand regulatory expectations and regulators to become more familiar with the technical aspects of neural networks, creating a framework where consumer protection and innovation can coexist in a balanced manner.

Regulators should be actively involved in the process of developing and deploying these technologies, working hand in hand with developers to ensure that artificial intelligence systems operate in a transparent and equitable manner. This involves defining specific guidelines for the treatment of data, the interpretability of models and the prevention of potential biases. Collaboration will also foster an environment of knowledge exchange, where best practices can be shared and adopted by different actors in the sector.

In addition, financial institutions have a key role to play in ensuring that neural networks are applied responsibly. To this end, they need to commit not only to comply with established regulations, but also to go further by establishing internal policies that ensure fair and ethical use of the technology. Collaboration among all these players will not only facilitate the adoption of neural networks, but will also contribute to establishing a safer, more transparent and fairer financial environment for all involved, thus strengthening public confidence in the technology.

Bibliographic References

Aldabas-Rubira, J. (2002). *Redes Neuronales y su Aplicación en Predicciones de Consumo.* Editorial Científica de Tecnología.

Basogain Olabe, X. (2014). *Artificial Neural Networks and their Applications.* School of Engineering of Bilbao, EHU.

Del Carpio Gallegos, A. (2005). *Neural networks for credit management.* Editorial Nombre.

Fernández, R., & Torres, L. (2024). *Supervised and unsupervised learning in credit risk assessment.* Editorial Nombre.

García López, H. (2024). *Data cleaning in preprocessing for credit risk management.* Editorial Nombre.

Guerrero, W. A., Camacho-Galindo, S., Guerrero-Martin, L. E., Arévalo, J. C., de Freitas, P. P., Gómes, V. J. C., Fernandes, F. A. S., & Guerrero- Martin, C. A. (2024). Impact of artificial intelligence on financial decision making: opportunities and challenges for business leaders. DYNA, 91(233), 168-177.

Hebb, D. O. (1949). The Organization of Behavior: A Neuropsychological Theory. Wiley.

Hernández, G., & Vásquez, P. (2023). *Unsupervised learning in credit management.* Editorial Nombre.

Herrera Ortega, R. (2023). *Ethical Considerations in the Use of Neural Networks for Credit Management.* Editorial Universitaria.

Lundberg, S. M., & Lee, S.-I. (2017). *A Unified Approach to Interpreting Model Predictions.* Advances in Neural Information Processing Systems (NIPS).

McCulloch, W. S., & Pitts, W. H. (1943). A Logical Calculus of the Ideas Immanent in Nervous Activity. Bulletin of Mathematical Biophysics, 5, 115-133.

Melchor Pérez, S., et al. (2024). *The output layer in neural networks in the financial context.* Editorial Nombre.

Méndez Araya, VE (2009). Study of the application of neural networks in credit risk assessment (Final report of Civil Engineering in

Computer Science). Pontificia Universidad Católica de Valparaíso.

Montalván, F. (2019). *Neural networks applied to supervised learning.* Editorial Nombre.

Perales Paz, J. (2024). *Structure and operation of neural networks for credit risk management.* Editorial Nombre.

Pérez Ramírez, F O., & Fernández Castaño, H. (2007). *Neural Networks and Credit Risk Assessment.* Revista Ingenierías Universidad de Medellín, 6(10), 77-91.

Pérez Ramírez, L., & Fernández Castaño, R. (2007). *Error backpropagation in neural networks*. Editorial Nombre.

Ramírez, J. A., & Chacón, M. I. (2011). Artificial neural networks for image processing, a review of the last decade. Journal of Electrical, Electronic and Computer Engineering (RIEE&C), 9(1), 7-16.

Ribeiro, M. T., Singh, S., & Guestrin, C. (2016). *"Why Should I Trust You?" Explaining the Predictions of Any Classifier.* Proceedings of the 22nd ACM SIGKDD International Conference on Knowledge Discovery and Data Mining.

Rodríguez, M., & Martínez, C. (2023). *Normalization and standardization of data in the credit context.* Editorial Nombre.

Ruiz, L., & Basualdo, A. (2001). *Hebb's Rule and its Application in Modern Neural Networks*. Editorial Científica.

Rumelhart, D. E., Hinton, G. E., & Williams, R. J. (1986). Learning Representations by Back-Propagating Errors. Nature, 323, 533-536.

Sosa Sierra, M. del C. (2007). Artificial intelligence in corporate financial management. Thought & Management, (23), 153-186.

Toro Ocampo, A., Mejía Giraldo, J., & Salazar Isaza, P. (2004). Title of the book or article. Publisher or source of publication.

Van Greuning, H., & Brajovic Bratanovic, S. (2009). Bank risk analysis: A framework for assessing corporate governance and risk management (3ª ed.). World Bank. Ladino Becerra, I. C. (2014). *Comparison of Credit Risk Models: Logistic Models and Neural Networks.* Pontificia Universidad Javeriana.

Villamil Bahamón, R. (2013). *Predictive Neural Model for Credit Risk Assessment.* National University of Colombia.

Werbos, P. (1974). Beyond Regression: New Tools for Prediction and Analysis in the Behavioral Sciences. Ph.D. Dissertation, Harvard University.

Printed by Books on Demand GmbH, Norderstedt / Germany